How To Solder Electronics

Complete Guide To Solder Electronics And Join Wires Together

Copyright@2023

Cary Leighton

Table Of Content

CHAPTER ONE 3
   Instructional Guide to Solder Electronics 3
   Section 1-Acquire All Of The Required Tools And Supplies 4
   Section 2- Soldering Each Individual Component 15
   Section 3-Soldering Well 23

CHAPTER TWO 33
   Instructional Guide To Solder Wires Together 33
   Section 1-Splicing Your Wires 34
   Section 2-Putting The Solder On The Joints 44
   Section 3- Sealing The Connection Together 52

CHAPTER ONE

Instructional Guide to Solder Electronics

Learning to solder components with through-holes is a skill that is vital for every amateur electronic hobbyist as well as any electronic expert. You will be able to acquire the knowledge and experience necessary to start soldering electrical components in the correct manner.

Section 1-Acquire All Of The Required Tools And Supplies

1-Make sure the soldering iron you're using has an adjustable temperature control.

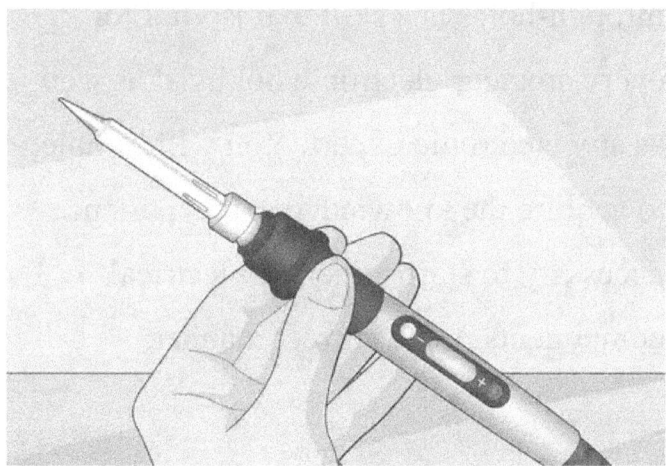

The most effective soldering irons are those that are resistant to electrostatic discharge (ESD), have temperature controls, and a high power output. These soldering irons are ideal for soldering electrical components onto printed circuit boards. You can solder for hours with them, and they are useful for putting together complicated amateur radio projects. For more straightforward kits, a low-cost pencil iron should do the job just fine.

- When working on smaller projects, use a soldering iron with a power setting of 25 watts; when working on bigger projects with heavy wiring, increase the power setting to 100 watts
- It is recommended that irons with changeable temperature settings be made accessible, as this will provide the most secure handling of the boards. The temperature of the tip may be adjusted to correspond with the scale of the work being done.

2-Solder wire composed of a suitable alloy should be used.

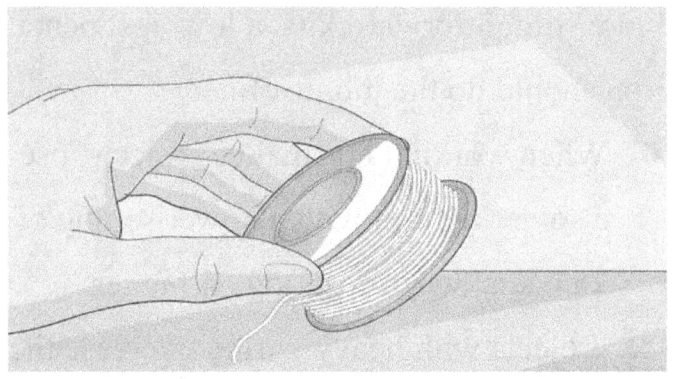

The ratio of tin to lead in the most popular solder alloy used in electronic components is commonly written as 60/40 (SN/Pb), however the real ratio that has the lowest melting temperature is 63/37. If you are new to soldering, it is advised that you do this, despite the fact that because to the lead concentration, it is fairly risky. You are required to utilize the appropriate ventilation, a suitable breathing mask, or soldering equipment that is equipped with a vacuum attachment.

- Solder with a composition of 60/40 becomes malleable at 361 degrees Fahrenheit (183 degrees Celsius), but it doesn't melt until 370 degrees Fahrenheit (188 degrees Celsius), which means that it might be challenging to deal with for those who are just starting out. Solder with a 63/37 composition may be used instead since it can be melted at 361 degrees Fahrenheit (183 degrees Celsius).
- In recent years, as a result of the RoHS regulatory initiative, the use of a variety of lead-free alloys has been obligatory. These have a greater temperature requirement for the soldering process and do not "wet" as well as tin-lead alloys. Although they are less dangerous, they are also more difficult to understand. The most

typical ratio is 96.5 percent tin to 3.5 percent silver, which will provide a joint that has lower electrical resistance than an alloy made of tin and lead. In actuality, this is not a cause to utilize it; the primary motivating element is the concern for people's safety. Solder that is nearly entirely composed of tin may also be purchased,

3-When working with electricity, you should make use of a solder wire that has a flux core.

Ensure that the flux being utilized is electrically compatible. Solder flux typically used in plumbing is not electrically compatible. When preparing surfaces for soldering, flux is a substance (rosin or a version for electrical work) that is applied to the area. It is imperative that any contaminants, such as grime, grease, and the like, be removed before soldering. Although extremely tiny surface mounting or automated soldering may employ alternate methods, the most practical approach is to include the flux inside the solder wire itself. This ensures that flux is automatically supplied to the surfaces that are being soldered.

- When working with electricity or electronics, one may choose from a wide variety of fluxes that are commercially accessible. RMA, RA, and water-soluble fluxes are the three most common types. The more active a flux is, the more vital it is that it does not stay after soldering. Otherwise, continued chemical activity might jeopardize or harm the performance of the electrical or electronic equipment. In particular, fluxes that are water-soluble need to be eliminated.

- Following the process of soldering, rosins leave behind a dark, sticky residue that, ideally, does not corrode and does not carry electricity. Isopropyl alcohol or a specially designed rosin removal solution may also be used for cleaning purposes. Rosin removal products are also available.
- After the soldering process, no-clean flux will leave behind a clean residue that will not corrode and will not

conduct electricity. This flux is intended to be left on the solder junction as well as the surfaces around it.

- The water-soluble flux often has a greater activity level, which results in the formation of a residue that needs to be removed using water. If the residue is not properly wiped up after usage, it is corrosive and may potentially cause harm to the circuit board or its components.

4-Obtain the required board as well as the components.

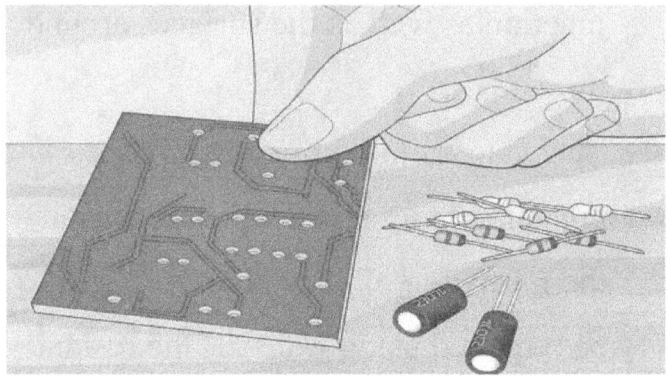

The majority of soldering in the electrical industry is done using "through-hole" components. These are components whose leads are put into holes in printed circuit boards (PCBs) and then soldered to a pad of metal plating (a PCB trace) surrounding the hole. If the interior of the hole is not "plated through," then the inserted lead serves as the electrical connection between the traces on the top and bottom of the printed circuit board (PCB). However, if the hole is "plated through," then the opposite is true. In the final scenario, it is often essential to solder the lead on both sides of the component.

- Soldering various electrical components, such as cables or lugs, requires somewhat different approaches, although the essential concepts of operating the solder and iron remain the same. Other sections of this book provides a description of these methods. Note, however, that a dependable mechanical connection must exist between the lugs and any other unsupported soldering sites before soldering can take place. Solder joints are only useful for making electrical connections with extremely low resistance; they do not contribute to the mechanical strength or resistance to vibration of a joint.

5-Obtain a clamp to use for holding the various components.

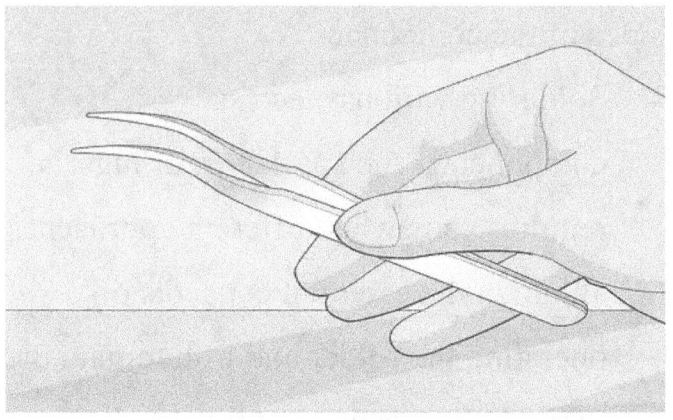

Because electrical components are often relatively tiny, you will need tweezers, tongs, or needle-nosed pliers to hold them in place while you use the soldering iron and work with the solder. Finding the ideal balance might be challenging.

- When soldering components to a circuit board, it is ideal to have some form of stand or clamp to keep the board in place while you work.

Section 2- Soldering Each Individual Component

1-Get the individual components ready to be soldered.

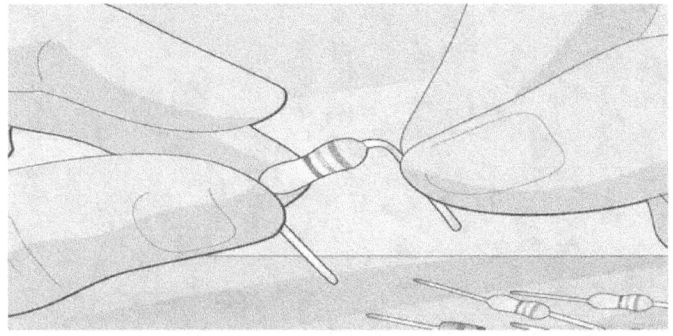

Determine which component is the right one by paying close attention to both its type and value. When it comes to resistors, be sure to verify their color code. In the event that it is essential to do so, appropriately bend the leads, taking care not to exceed the stress specifications in any way (for example, by bending them too sharply), and clinch the leads so that they fit the board.

2-**Always use great caution, and solder in settings that are designed for the activity**.

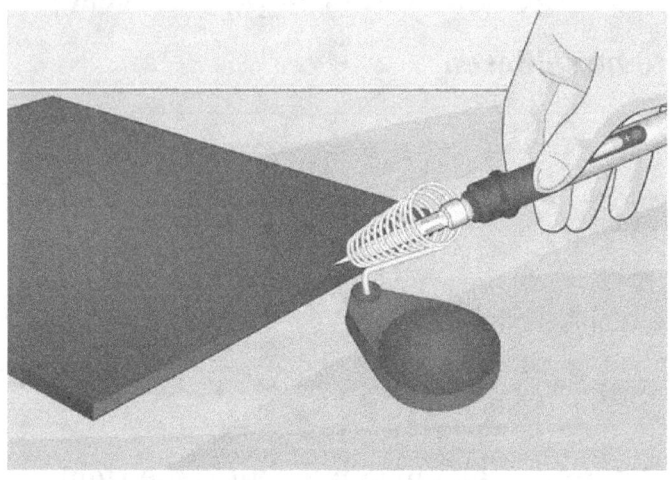

Always solder in a well-ventilated environment while wearing protective gear for your eyes and breathing. When the iron is turned on but not being used, be careful to put it in a secure location (using a stand or holder that is fireproof). Burning paper, plastic, or even your workbench may easily cause irons to ignite flames when they come into contact with them. When protecting the region, a heat mat or board should always be used.

- If you don't leave a distance of at least 7–12 inches (18–30 cm) between the electrical components and your face, solder pieces and hot flux might potentially go into your eyes. Wearing safety glasses is a very prudent preventative measure. Solder that is melting has the potential to spatter and is inherently unpredictable.

3-Apply "tin" on the tip of the soldering iron.

On the tip of the soldering iron, carefully melt a little dab of solder. Tinning is a method that helps to increase heat transfer from the iron to the lead and pad, which in turn protects the board from being subjected to high heat for an extended period of time.

- Place the tip, which still contains the glob, carefully onto the interface between the lead and the pad. It is necessary for the point or blob to make contact with both the lead and the pad.
- The non-metallic region of the printed circuit board (PCB) should not come into contact with the tip of the soldering iron, regardless of whether it is made of fiberglass (which is fairly common) or another material. This region is vulnerable to harm if the temperature becomes too high.

4-Feed the solder wire onto the interface that is comprised of the pad and the lead.

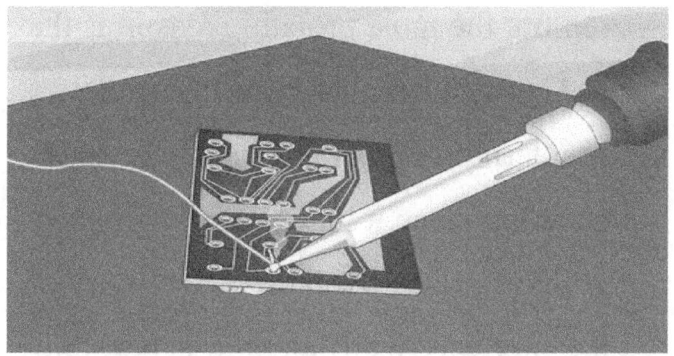

After melting onto the junction, the flux from the solder wire is only active for a very limited period of time, at most. It is burnt away gradually, which may be seen as smoke coming from the joint, and as this process continues, it loses its potency. It is necessary to provide sufficient heat to both the component lead and the pad in order to melt the solder into the connecting point. It is expected that the liquid solder would "cling" to the pad and lead together due to surface tension. *Wetting* is the term that most people use to refer to this process.

- If the solder does not melt onto the region, the most probable reason is that not enough heat has been transmitted to it; alternatively, the surface may need cleaning since it may be contaminated with oil or dirt. Because the activity of the flux was insufficient, it is possible that an external flux will be required. Prior to soldering, it is possible that the surfaces may need careful cleaning.
- Use extreme caution since sandpaper will almost always be too abrasive, and steel wool, although being less physically abrasive, will add microscopic particles of conductive metal. This will most likely result in inadvertent shorts and electrical behavior that is undesirable.

5-When all of the surfaces have been wetted, you should stop supplying fresh solder.

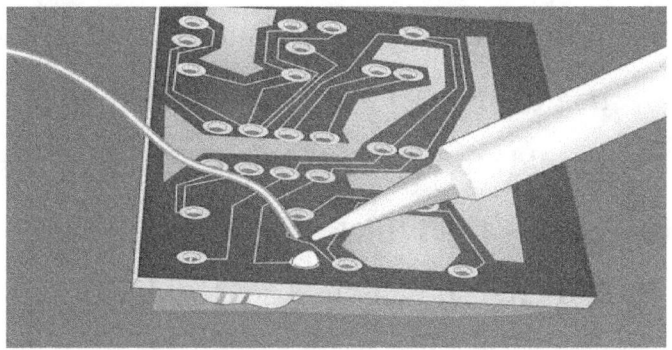

You should stop adding extra solder after the gaps have been filled in and the surfaces have been wetted. The majority of junctions should need little more than a drop or two of solder at most, however the amount required may vary somewhat depending on the component being joined. The appropriate quantity of solder is determined by the following criteria:

- On plated-PCBs, you need to stop feeding when you can see a solid concave fillet around the joint.
- When working with printed circuit boards (PCBs) that are not plated, you should stop feeding the solder when it creates a flat fillet.
- If there is not enough solder, the junction will be uneven and concave, whereas if there is too much solder, a bulbous joint with a convex shape (i.e., blob-like) will be formed. Both of these are outward signs that the solder junction was not properly completed.

Section 3-Soldering Well

1-You should move rapidly.

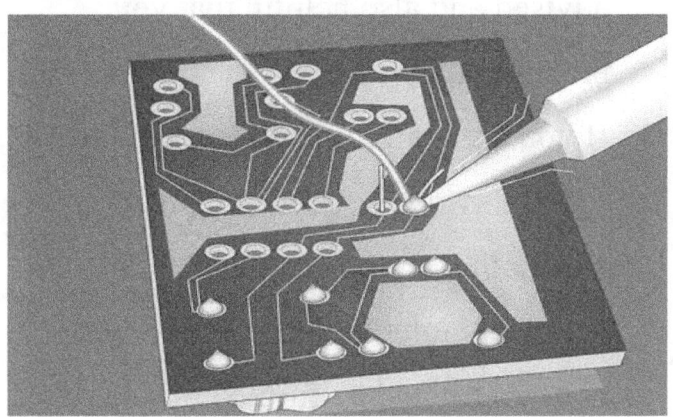

Unfortunately, it is not difficult at all to ruin a component or the board by applying excessive heat to it. Moving quickly, on the other hand, will allow you to preserve the integrity of the circuit board and the components on it. Keeping an adjacent finger on the board might be helpful in detecting excessive heat.

- Always err on the side of caution and choose irons with a lower power rating than you believe you will need. For the majority of electronic work, a 30-watt iron should be sufficient. It is strongly advised and also helpful that you practice soldering.

- If you are working with a circuit board that has two sides, check both sides to ensure that the solder joints are secure. A healthy joint will have a glossy appearance and be formed like a cone. If the joint seems icy and lifeless, then it most likely is a cold joint.

2-To safeguard critical components, give some thought to using heat sinks.

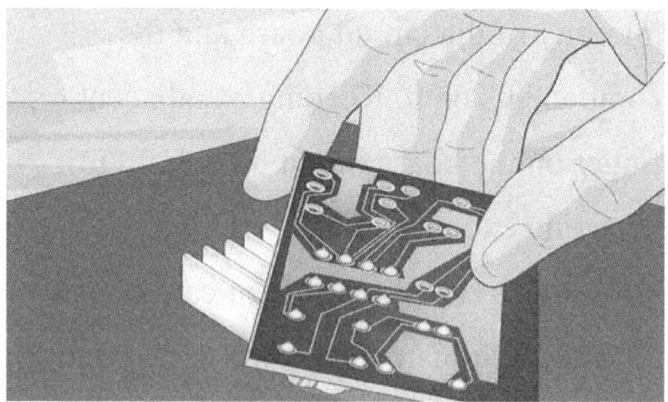

Some components, such as diodes and transistors, are very prone to heat damage and need the use of a tiny aluminum heat sink that is hooked on to the printed circuit board's (PCB) opposite leads. Electronics supply stores often carry the little aluminum heat sinks that are used in electrical devices. Small hemostats are another option that may be employed.

3-You should train yourself to detect when there is an enough amount of solder present.

Following an appropriate coating of solder, the solder will have a glossy appearance rather than a dull one. The use of visible indicators is the most reliable method for determining whether or not a solder junction is sound. It is more important for the solder to melt onto the surface of the electrical components or the PCB traces than it is for it to melt on the tip of the soldering iron. In this manner, a strong connection will be formed between the surface of the metal and the solder after it has cooled.

- The solder joint should equally cover the surface of the component. There should not be too much solder, as this might cause it to form a glob, nor should there be too little solder, as this could result in the surface not being entirely coated.

4-Ensure that the soldering iron is always clean.

It is possible for the tip of the soldering iron to get contaminated by burnt flux, rosin from the center of the solder, or plastic sheaths from wires. These pollutants stop a proper connection from forming between the electrical components and prevent them from working properly. This is not desired because it increases the electrical resistance and decreases the mechanical strength of the solder junction.

Neither of these outcomes is good. A clean tip is shining all the way around and free of any charred gunk that may be present.
- After you have finished soldering each component, clean the iron well. To ensure a thorough cleaning, you may use either bronze (or brass) wool or a moist sponge.

5-Before relocating the components, wait until the solder has fully cooled down.

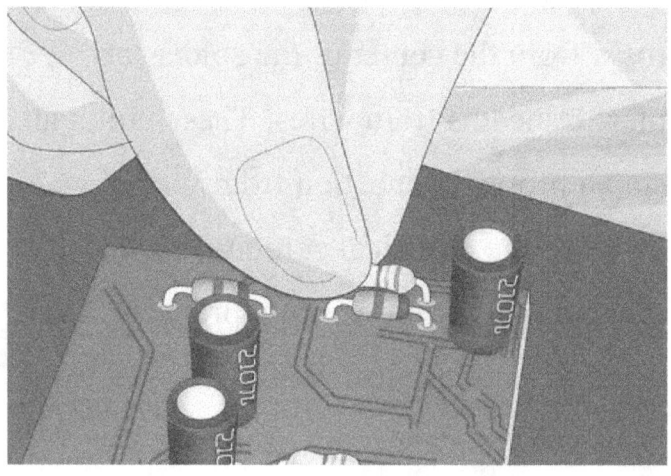

Solder is malleable for some time, and there is not much of a telltale sign to indicate when the mushy period is finally over. This cooling should only take a few seconds in the majority of electrical scenarios; but, big components have a greater mass, making it more difficult to heat them enough to solder them, and also requiring considerably more time to cool before they become solid.

- If the components are too hot to handle safely, you may use needle-nose pliers or a gadget called helpful hands, which consists of two alligator clips coupled to a tiny articulated platform. If you keep a close eye on it, the solder that is cooling will eventually settle in front of your own eyes.

6-Get your feet wet with discarded components.

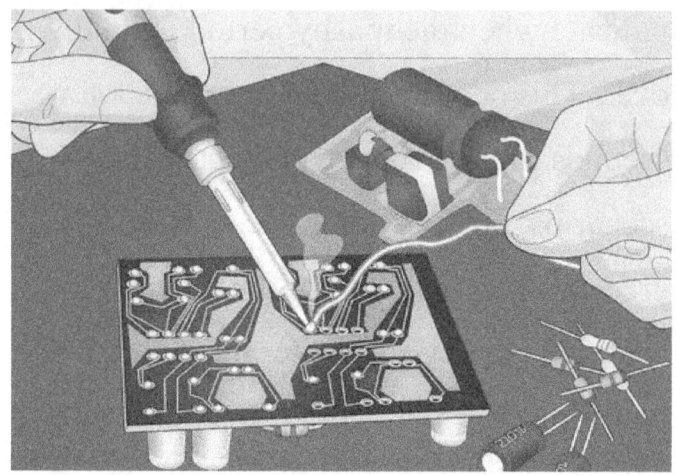

Before attempting to solder anything significant for the first time, it is essential to gain experience by first working with less significant materials. You should practice on some garbage components obtained from an old radio or something like.

- No one, not even those who work in the field, is flawless. It is quite OK to rework some of the soldering that you have already done; the term "rework" is used formally in the industry. It's going to save you time in the future when you are troubleshooting.

CHAPTER TWO

Instructional Guide To Solder Wires Together

Soldering is a technique that requires melting a minimal metal alloy over a joint or wire splice in order to hold two components together so that there is no possibility of their falling apart. If you wish to join two wires together, you may use solder to quickly create a connection that will hold up for a significant amount of time. To initiate the connection, begin by stripping the wires and winding them around each other in a clockwise direction. After that, you may proceed to solder the wires in place by melting the solder and applying it straight to the wires. After covering and waterproofing the exposed wires to make a seal around them, you are done!

Section 1-Splicing Your Wires

1-Remove a quarter of an inch (2.5 centimeters) of insulation from one end of each wire.

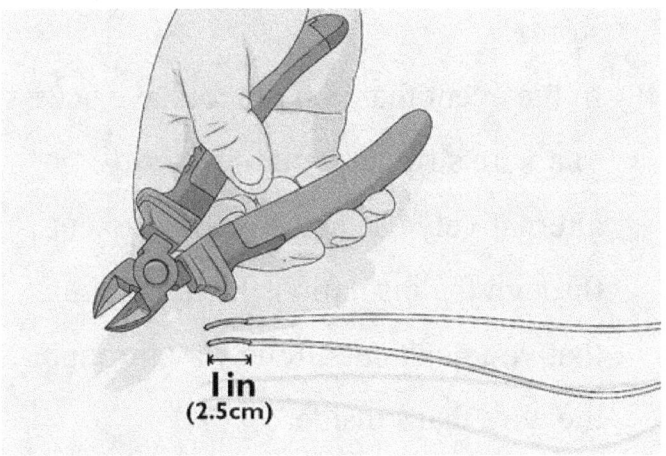

Place the jaws of a wire stripper an inch and a half (2.5 centimeters) from the end of one of the wires that you are going to splice together. When you want to remove the insulation from the wire, squeeze the handles together tightly and then draw the jaws toward the end of the wire. It is necessary for you to repeat the technique on the end of the second wire that you are splicing.

- Wire strippers are available at most home improvement and hardware stores.
- In the event that you do not have access to a wire stripper, you may alternatively use a utility knife to cut through the insulation. Just make sure that you don't cut all the way through the wire that's inside.
- If strands of a stranded wire are accidently broken off, the wire may cause a fuse to explode in the device it is attached to. Remove any further strands of insulation from the wire, and then attempt to strip it once again.

2-Attach a piece of heat-shrink tubing to one of the wires by sliding it onto the wire.

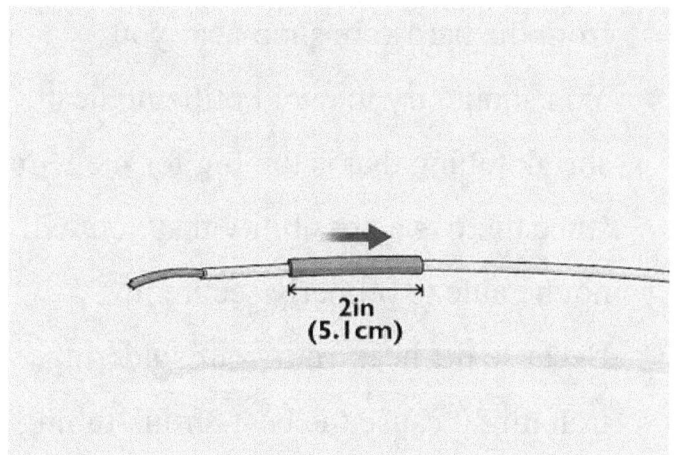

To make it easier to put the heat-shrink tubing onto the wire, you should get it at a gauge size that is one bigger than the wire you are using. To ensure that the piece of tubing you cut is long enough to subsequently cover the splice as well as portion of the insulation, cut it to a length of at least 2 inches (5.1 cm). Slide the heat-

shrink tubing onto one of the wires, and then move it a distance of at least 1 foot (30 cm) away from the wire's exposed end.
- You may purchase heat-shrink tubing from the hardware shop near you.
- You should try to avoid utilizing heat-shrink tubing that is too big for the wire, since there is a possibility that you will not be able to properly secure it.
- Because the heat from your soldering iron might cause the heat-shrink tubing to contract, you should not keep it close to the region where you are soldering it.

3-To join the two wires together, step three is to twist their respective ends together.

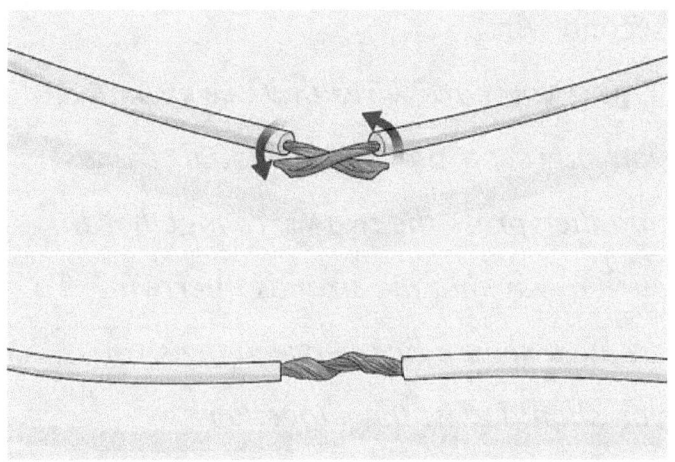

Put the centers of the exposed wires in line with one another so that they form an X shape. Make sure the connection is secure by bending one of the wires down and then firmly twisting it around the other wire as many times as you can. It is important to check that the end of the wire does not protrude above or point away from the splice, since this will result in a connection that is not as secure. To ensure that the splice is uniform on both sides, you will need to repeat the technique with the second wire.

Tip: *If you have wires that are stranded, you may separate the individual strands and then press the two wires together in such a way that the strands intertwine. To create a secure link between the strands, you should twist them together.*

Tip: *lineman pliers or Needle nose pliers should be used in the event that you need to straighten up the wires so that it is simpler to align them with one another. Take each wire in turn and work on it in little pieces at a time until you get it back to its straight state.*

4-Secure the wires with alligator clips so they won't get in the way of your job and secure them with the clips.

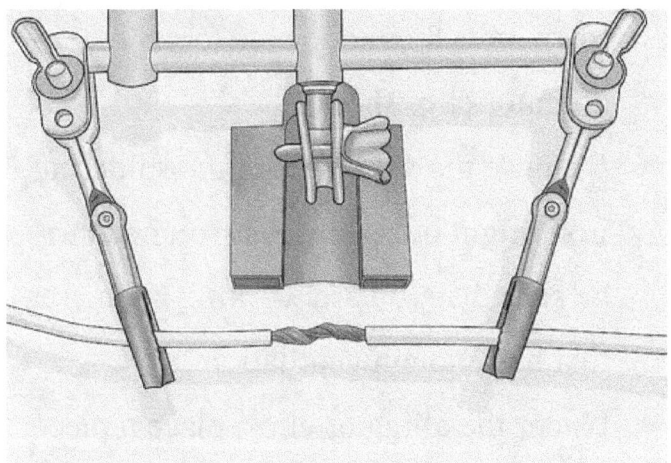

Alligator clips are little grips made of metal that are effective at preventing wires from moving about while they are being held in place by the alligator clip. Position the alligator clips so that the jaws are facing upward and vertically align them on a level work surface. Put each of the wires into its own alligator clip and then secure the alligator clips together such that the splice may be held off the work surface between them.

- If you go to the hardware shop in your town, you should be able to get alligator clips there.
- Because the vapors from the soldering iron might be hazardous, you need to be sure that you are working in an area that has enough ventilation.
- Under the alligator clips, place a piece of scrap metal or another material that will not catch fire to collect any solder that may leak.

5-Apply rosin flux to the spliced wire in order to improve the solder's ability to cling to the wire.

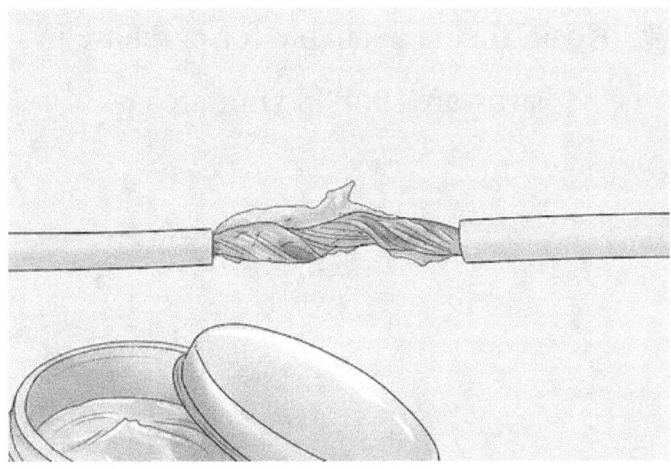

A rosin flux is a kind of substance that not only assists in cleaning the wires but also makes it possible for solder to adhere to them. You should put a little quantity of rosin flux, about the size of a bead, on your finger and then brush it over the exposed wires. Make an effort to apply the flux to the wires in an even and uniform manner so that a thin film is formed on them. Use

your finger or a piece of paper towel to remove any excess flux that may be on the wires.
- Rosin flux is available for purchase in the hardware shop in your area.

Section 2-Putting The Solder On The Joints

1-For the material that is the least difficult to work with, get 63/37 leaded solder.

Solder is often fabricated using a mixture of metals that are able to melt at a relatively low temperature, such as tin or lead. When heated to 361 degrees Fahrenheit (183 degrees Celsius), 63/37 solder, which is composed of 63% tin and 37% lead, immediately transforms from a solid state into a liquid state. If you are dealing with electronics, the easiest way to link the wires together is to use solder with a 63/37 melting point ratio.

- Consuming lead may expose you to a variety of health risks, thus it is important to properly wash your hands after working with lead in solder. Because you won't be working with the solder for very long, gloves aren't necessary, but you should consider putting them on nonetheless just in case.
- Lead-free solder is also available, however its manipulation could be trickier than traditional solder.
- Don't use silver solder since it's often reserved for plumbing and other pipe-related applications.

2-To prevent oxidation from occurring, melt some solder on the tip of your soldering iron.

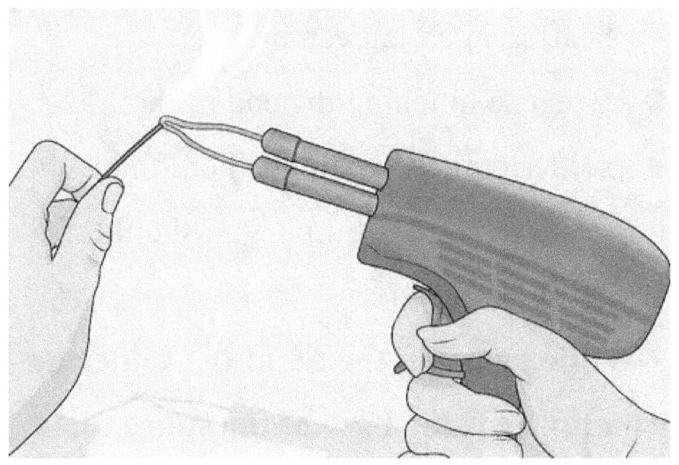

Put on a set of goggles or safety glasses to guard against damage to your vision. It should only take a few minutes, but you should start by turning on your soldering iron and letting it heat up entirely. Keep the tip of your solder directly on the tip of the iron so that it melts into a thin coating on the iron. Keep applying solder to the iron until it has a glossy look to it.

- This procedure, known as "tinning" the iron, prevents oxidation, which is one of the factors that might lead to the iron heating in an uneven manner.
- Because touching the end of the soldering iron when it is hot might result in serious burns, you should avoid doing so.

3-While pressing the tip of the soldering iron towards the base of the splice, the flux should be heated.

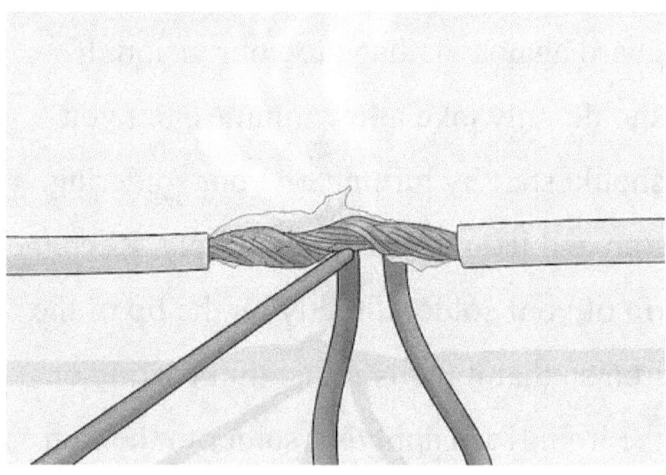

Maintain the heat of the soldering iron and position it so that it is touching the underside of the wire splice. Because of the heat that is transferred from the iron into the wires, the flux will change into a liquid state. After you see that the flux has begun to bubble, you may start applying solder to the splice.

- Wires with a thicker gauge may need more time to reach operating temperature compared to those with smaller gauges.
- Put on some old clothing that you won't mind being burned if you happen to contact the soldering iron or hot solder by mistake.

4-Move the point of the solder along the wire so that it melts into the wires.

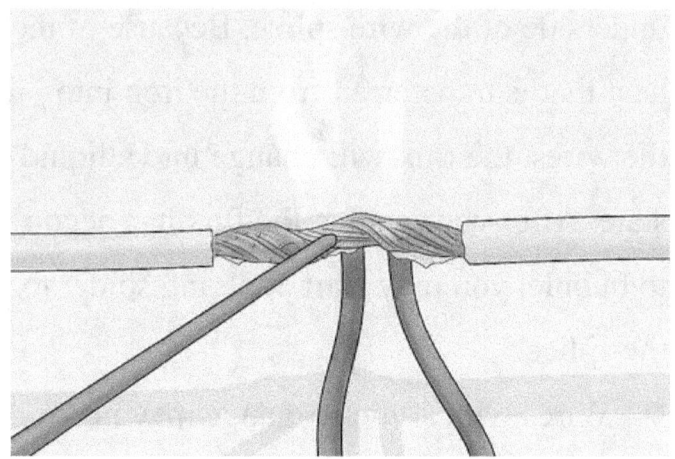

Maintain contact with the wire with the soldering iron so that it may continue to be heated. The end of the 63/37 solder should be tapped on top of the wire splice so that it melts down into the wires. Move the solder around the splice in a circular motion so that it may melt and spread out into the spaces between the wires. Continue to melt the solder until a thin coating of solder covers all of the exposed wire and then stop.

- It is important to avoid breathing the fumes that are produced by the solder since they might irritate your respiratory system and be detrimental to your health. To prevent the accumulation of fumes, you should do your work in a location that has enough ventilation.
- It is up to you whether or not you wish to cover your face with a mask, although doing so is not compulsory.

Caution: *When you apply the solder to the wires, do not touch the solder directly to the soldering iron because this generates "cold solder," which results in a connection that is not as reliable and may cause a fuse to blow.*

5-Wait one to two minutes for the solder to harden once it has been allowed to cool down.

When you are completed, move the soldering iron and solder away from the splice so that it may have a chance to cool down. It is important not to touch or disrupt the wire while it is drying since doing so might cause the connection between the two to become less secure. After approximately one to two minutes, the solder will begin to harden, at which point you will once again be able to handle it.

Section 3- Sealing The Connection Together

1-To ensure that the soldered wire is watertight, rub some silicone paste over it.

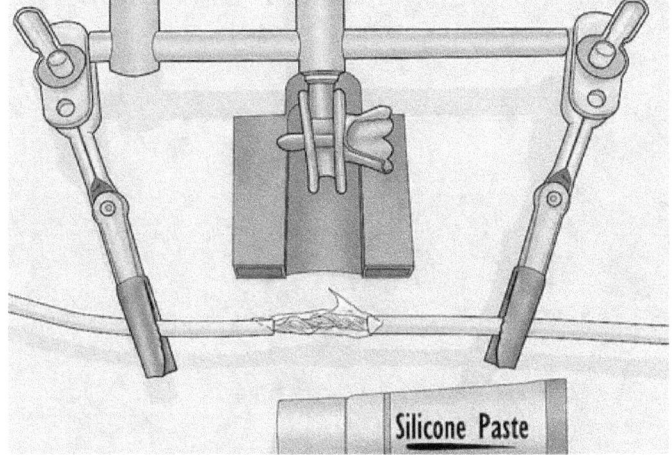

The use of silicone paste, which is often referred to as dielectric grease, protects the metal wires from rusting and ensures that the splice is totally watertight. Make use of a quantity of silicone paste equivalent to a bead, and distribute it all over the soldered wire using your finger. Make sure there is a

uniformly thin coating of the silicone paste on the wire so that it will remain protected.
- You should be able to get silicone paste at the hardware shop in your area.

2-Wrap the exposed wires with the heat-shrink tubing by sliding it over the wires.

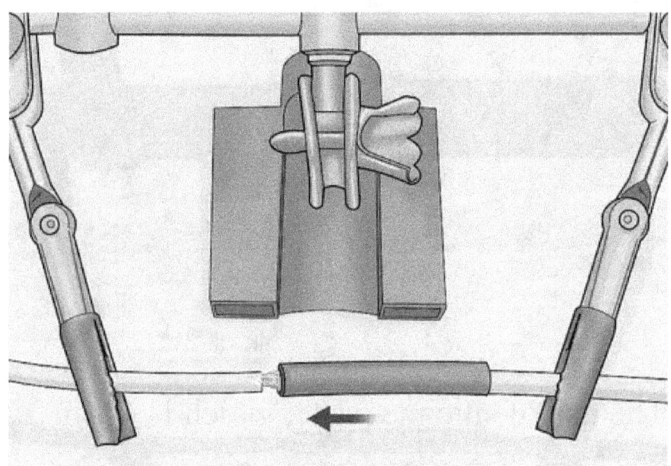

You should slip the heat-shrink tubing that you previously attached to the wire back over the soldered wire now that it has been removed. Be sure that the edges of the heat-shrink tubing reach over the insulation by at least a quarter of an inch (0.64 cm), so that there is no exposed wire visible through the tubing.

- It is not a problem if some of the silicone seeps out from below the heat-shrink tubing since there will still be enough of it on the wires to keep them protected.

3-Shrink the tubing that is placed over the soldered wires by using a heat gun.

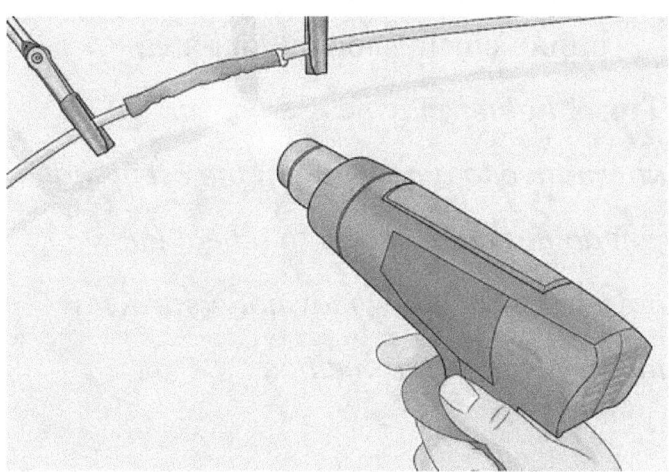

Keep the heat gun at a distance of about 4 to 5 inches (10 to 13 cm) from the tube.

Start applying heat to the middle of the tube after adjusting the heat gun's setting to the lowest possible level. Work your way around the wire's full circle, heating it from the middle outwards so that extra silicone paste seeps out of the sides as you go. You may remove the heat from the heat-shrink tubing after it has been pulled tightly around the wire.

- You may get a heat gun at the home improvement shop in your area.

Tip: *A lighter may be used as an alternative to a heat gun in the event that you do not have access to a heat gun; however, the tubing may not contract as uniformly with this method.*

4-Using a paper towel, remove any excess silicone paste from the surface.

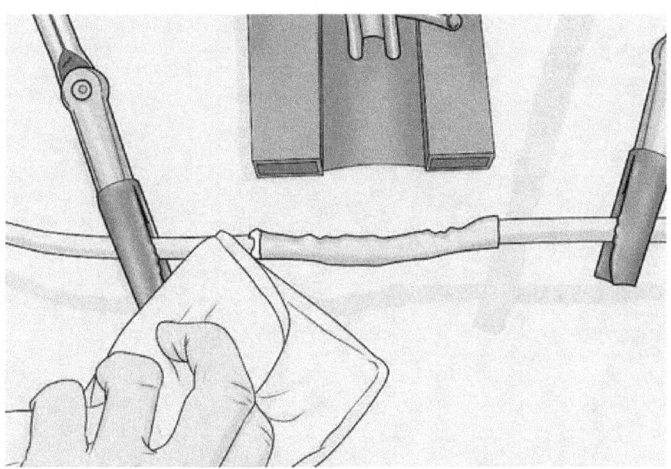

As the tube shrinks, there will a portion of the silicone paste that escapes from the side openings. When the wire and tubing are no longer hot to be touch, clean the wires by wiping the silicone off of them with a piece of paper towel so that they are free of residue. When you have completed removing the silicone paste, your wires will be complete.

www.ingramcontent.com/pod-product-compliance
Lightning Source LLC
Chambersburg PA
CBHW050312220526
45465CB00005B/1961